I0471173

Massive Leak of Liquefied Chlorine Gas
Henderson, Nevada

Reported by: J. Gordon Routley

This is Report 052 of the Major Fires Investigation Project conducted by TriData Corporation under contract EMW-90-C-3338 to the United States Fire Administration, Federal Emergency Management Agency.

FEMA

Department of Homeland Security
United States Fire Administration
National Fire Data Center

U.S. Fire Administration Fire Investigations Program

The U.S. Fire Administration develops reports on selected major fires throughout the country. The fires usually involve multiple deaths or a large loss of property. But the primary criterion for deciding to do a report is whether it will result in significant "lessons learned." In some cases these lessons bring to light new knowledge about fire--the effect of building construction or contents, human behavior in fire, etc. In other cases, the lessons are not new but are serious enough to highlight once again, with yet another fire tragedy report. In some cases, special reports are developed to discuss events, drills, or new technologies which are of interest to the fire service.

The reports are sent to fire magazines and are distributed at National and Regional fire meetings. The International Association of Fire Chiefs assists the USFA in disseminating the findings throughout the fire service. On a continuing basis the reports are available on request from the USFA; announcements of their availability are published widely in fire journals and newsletters.

This body of work provides detailed information on the nature of the fire problem for policymakers who must decide on allocations of resources between fire and other pressing problems, and within the fire service to improve codes and code enforcement, training, public fire education, building technology, and other related areas.

The Fire Administration, which has no regulatory authority, sends an experienced fire investigator into a community after a major incident only after having conferred with the local fire authorities to insure that the assistance and presence of the USFA would be supportive and would in no way interfere with any review of the incident they are themselves conducting. The intent is not to arrive during the event or even immediately after, but rather after the dust settles, so that a complete and objective review of all the important aspects of the incident can be made. Local authorities review the USFA's report while it is in draft. The USFA investigator or team is available to local authorities should they wish to request technical assistance for their own investigation.

This report and its recommendations were developed by USFA staff and by TriData Corporation, Arlington, Virginia, its staff and consultants, who are under contract to assist the USFA in carrying out the Fire Reports Program.

The USFA greatly appreciates the cooperation received from Fire Chief William Bunker of the Clark County Fire Department, as well as Assistant Fire Chief Tom Alexander and Public Information Officer Robert Leinbach of his staff, and also Battalion Chief William Kourim of the City of Henderson Fire Department. Information was also provided by Cathleen Faulx, Emergency Management Coordinator of the Clark County Emergency Preparedness Office and Plant Manager Clifford H. Barr of Pioneer Chlor Alkali Company, Inc.

For additional copies of this report write to the U.S. Fire Administration, 16825 South Seton Avenue, Emmitsburg, Maryland 21727. The report is available on the Administration's Web site at http://www.usfa.dhs.gov/

U.S. Fire Administration

Mission Statement

As an entity of the Department of Homeland Security, the mission of the USFA is to reduce life and economic losses due to fire and related emergencies, through leadership, advocacy, coordination, and support. We serve the Nation independently, in coordination with other Federal agencies, and in partnership with fire protection and emergency service communities. With a commitment to excellence, we provide public education, training, technology, and data initiatives.

TABLE OF CONTENTS

Massive Leak of Liquefied Chlorine Gas
Henderson, Nevada
May 6, 1991

Local Contacts: William Bunker, Fire Chief
Tom Alexander, Assistant Fire Chief
Robert Leinbach, Public Information Officer
Clark County Fire Department
575 East Flamingo Road
Las Vegas, Nevada 89119
(702) 455-7311

Cathleen Faulx, Emergency Management Coordinator
Clark County Emergency Preparedness
225 Bridger Avenue
Las Vegas, Nevada 89155

William Kourim, Battalion Chief
City of Henderson Fire Department
239 Water Street
Henderson, Nevada 89015
(702) 565-2307

Clifford H. Barr, Plant Manager
Pioneer Chlor Alkali Company, Inc.
8000 Lake Mead Drive
Henderson, Nevada 89015
(702) 565-8781

OVERVIEW

A massive leak of liquefied chlorine gas created a dangerous cloud of poison gas over the city of Henderson, Nevada, in the early morning hours of May 6, 1991. Over 200 persons were examined at a local hospital for respiratory distress caused by inhalation of the chlorine and approximately 30 were admitted for treatment. Approximately 700 individuals were taken to shelters. It is estimated that from 2,000 to 7,000 individuals were taken elsewhere.

SUMMARY OF KEY ISSUES

Issues	Comments
Situation	Chlorine release caused by leak of brine from heat exchanger mixing with liquefied gas. Mixture created corrosive acid which ate through pipes when product was transferred from storage tank. Leak increased as acid ate larger hole in pipe.
Delayed Alarm	Plant employees believed they could contain the leak. Fire department notified by passer-by who was unsure of source. Response delayed until second call provided additional information. Further delay caused by long response distance and several possible sources to check for hazardous materials (Hazmat) release.
Jurisdiction	Plant is located in Clark County island surrounded by the City of Henderson. Population at risk primarily in city. Agencies work together effectively.
Injuries	Firefighters and plant personnel overcome when chlorine cloud moved in unexpected direction. Command post had to be relocated three times to avoid moving cloud. Some residents exposed during evacuation; over 200 examined at hospitals; 30 admitted.
Evacuation	Citizens evacuated as leak continued to expand and control efforts proved unsuccessful. Approximately 700 people taken to shelters; 2,000 to 7,000 taken elsewhere. Police officers assisting with evacuation and traffic control exposed to gas cloud.
Control Measures	Corroded valve allowed product to flow. Fire department Hazmat team and plant personnel entered together to stop flow. First attempt to stop flow by inserting blank flange was unsuccessful; Teflon-coated plate had to be used because of rapid corrosion of steel.
Extent of Leak	Approximately 70 tons of chlorine escaped. Cloud dissipated with morning heat and winds.

LOCATION

The Pioneer Chlor Alkali facility is located in an industrial area, approximately 10 miles southeast of Las Vegas, Nevada. It is one of several chemicals and materials processing facilities that are located in the Basic Management Inc. complex. When the BMI complex was established, during World War II, it was located in the desert, several miles from any existing populated areas. Henderson was established nearby as a support community for the industries that were located in the complex. The Las Vegas metropolitan area has experienced rapid growth during the last decade and the City of Henderson has become a heavily populated suburb with more than 60,000 residents. The BMI complex currently occupies an unincorporated "island" under the jurisdiction of Clark County and is almost completely surrounded by the incorporated City of Henderson.

Public concern with the materials that are produced and stored in the BMI complex has created pressure to relocate the complex, away from populated areas. Several Hazmat incidents have occurred in the immediate area, including an explosion of ammonium perchlorate at an adjacent facility in 1988, which resulted in two deaths and 372 injuries that included 15 firefighters. The most recent incident occurred while the Nevada State Legislature was considering a bill to require the complex to be relocated to an isolated area, approximately 15 miles north of Las Vegas. (For a detailed description of the earlier incident, refer to Report 021 of this series, "Fire and Explosions at Rocket Fuel Plant, Henderson, Nevada.")

ORIGIN OF THE LEAK

Pioneer Chlor Alkali produces chlorine gas through the electrolysis of sodium chloride (table salt). The chlorine is used primarily for water treatment and is shipped to clients in railroad tank cars and highway tank trucks. By-products of the process, including hydrogen gas and sodium hydroxide, are used by other companies in the BMI complex. The chlorine is compressed and stored as a liquefied gas in a bank of eight storage tanks, each of 150-ton capacity. The system has a storage capacity of approximately 1,200 tons, although company officials indicate that their policy is to limit the storage on hand to 300 tons or less.

Chlorine (Cl_2)

Molecular Weight	70.9
Boiling Temperature	20 degrees Fahrenheit
Immediate Danger to Life or Health	30 parts per million (ppm)
Short Term Exposure Limit	3 ppm
Expansion Ratio (Liquid to Gas)	450-500

The liquefied chlorine is stored in tanks at a pressure of approximately 50 pounds per square inch (psi) and temperature of approximately 30 degrees Fahrenheit. The chlorine is dried in the process of compression and, when dry, is noncorrosive. If it is contaminated by water or water vapor, hydrochloric acid is produced and the resulting product is extremely corrosive. It is believed that a tube failure in a heat exchanger allowed water to mix with the chlorine going into one of the storage tanks. When workers began to transfer the contents of that tank to a railroad tank car, the corrosive liquid began to rapidly deteriorate the steel piping system.

The leak was first detected by automatic monitoring equipment, located near the storage tanks, at approximately 0110 hours. Employees responding to the alarm found a pinhole-size leak in a 2-inch elbow, located on a catwalk level approximately 10 feet above ground. The leak was a few feet beyond the valve on the discharge side of the pump which was used to transfer the liquid from the storage tanks to the rail car loading rack.

Attempts were made by plant personnel to stop the flow and patch the leak. The pump was shut down and the discharge valve was closed to stop the flow from the storage tanks to the leaking pipe. Management personnel were notified and members of the company's emergency team were called to respond to the plant. At this time the leak was considered to be minor and employees believed that it could be controlled without causing a major hazard to the plant or the surrounding area.

Plant employees were considered to be proficient in handling situations of this type. The Chlorine Institute, a trade association of companies involved in the manufacturing, distribution, and use of chlorine and related products, coordinates a system of mutual-aid emergency response teams. These CHLOREP teams are made up of member company employees who are trained and equipped to respond to emergency incidents involving chlorine. Pioneer Chlor Alkali operates the CHLOREP team for Southern Nevada and surrounding areas and the emergency response team equipment is stored within its facility. Joint training exercises had been conducted with the Clark County Fire Department Hazmat Team and other area fire departments.

EMERGENCY RESPONSE

At approximately 0150 hours a citizen notified Henderson Police Department that she had encountered a strong offensive odor when passing near the BMI complex on a major highway. The highway passes within one half mile of the complex, and the caller was concerned that there could be a release coming from one of the occupancies. The call was relayed to the Las Vegas City Fire Department Communications Center, which provides communications for the Clark County Fire Department. Since reports of odors in the area are a frequent occurrence, the Clark County Battalion Chief was notified, and he made the decision to wait for a more positive report of an incident before responding. The communications personnel began to call the industries in the area to ask if any of them had a problem.

At approximately 0200 hours a second call was received, this time by the Fire Communications Center, reporting a strong odor in the area. A full first alarm assignment was dispatched, including the Clark County Hazmat Team. The response time for the first units was 14 minutes due to the distance from the closest Clark County fire station. The first arriving units had some difficulty identifying which of several separate facilities was the source of the problem, until the odor was encountered. The odor was immediately recognized as chlorine, and the units were directed to the correct location.

Arriving at the gate of the Pioneer Chlor Alkali facility, the Clark County Battalion Chief found several employees who had been exposed to chlorine gas and were in need of medical attention. Employees reported that they were in the process of shutting down the plant and isolating the leak. They believed that the leak was successfully isolated and that they could handle the situation at that time.

Within a few minutes the atmosphere around the plant entrance became enveloped by the chlorine cloud and most of the plant employees donned their emergency escape respirators. The battalion chief requested assistance and ordered the area to be evacuated shortly before he and several other fire department members were overcome. All fire department and plant personnel evacuated to a location approximately one half mile uphill from the plant, where a command post was established. The Henderson Fire Department Battalion Chief, who had responded on mutual aid, assumed command and requested additional assistance, while the Clark County Battalion Chief and several other fire department members and plant employees were treated and transported to hospitals.

The Clark County Assistant Fire Chief responded from his home and assumed the duties of incident commander. He assigned the Henderson Battalion Chief, who was the most familiar with the area, as the operations chief in the incident command structure. The second arriving Clark County Battalion Chief was assigned to supervise the Hazmat group, due to his Hazmat team experience. Additional fire and medical units were directed to respond to a staging area adjacent to the command post. Liaison was also established with the Las Vegas Metropolitan and City of Henderson Police Departments at the command post.

STRATEGIC CONSIDERATIONS

Strategic decision making at this point focused on several factors:

1. Plant employees believed that the leak was isolated and would stop as soon as the residual chlorine had escaped from the piping system. The plant's chlorine production had been shut down and the valve between the storage tanks and the leaking valve was closed. This would leave only the liquid that was already in the piping system that could leak out. They believed that the cor-

rosive liquid was eating away at the elbow, but that not more than 1,000 lbs. of chlorine could escape. Based on this information, there was no point in risking personnel to take additional leak control actions.

2. The population immediately at risk included only the adjacent industrial facilities, most of which were shut down and evacuated. One facility could not be shut down and a minimal crew was left to operate it, while wearing self-contained breathing apparatus (SCBA). These employees were trained in Hazmat procedures.

3. The use of water fog to accelerate vaporization or to absorb the vapors would have required the commitment of personnel and equipment in a hazardous area. Long hoselines would have been required to place master stream appliances in positions where they might be effective, and the acid produced by the combination of water and chlorine could have caused a greater damage to plant facilities and equipment.

4. At this point the situation appeared to be stable. Based on the strategic considerations, a cautious "wait and see" approach was taken and all personnel were kept out of the immediate area of the leak. Hazmat group members with monitoring devices were deployed in an attempt to define the location and track the progress of the gas cloud. City of Henderson officials assembled at the command post and maintained an immediate awareness of the situation. Additional resources were staged and precautions were taken, in case the situation changed and evacuation became necessary.

EXPANDING GAS CLOUD

It was very difficult to accurately predict the size or travel of the gas cloud resulting from vaporization of the liquid pool on the ground. The flow rate of the leak could not be determined and the size of the resulting liquid pool could not be observed, since the storage tanks are surrounded by other processing equipment. The developing cloud could not be visually monitored due to the darkness and the location of the problem within the facility.

The heavier-than-air chlorine vapors have a tendency to move along the ground and concentrate in low spots. The BMI complex is located in a relatively high area and the terrain slopes down toward the north for several miles. A dry wash provided a low path for the heaviest concentration to migrate into a sparsely populated area north of the BMI complex. A growing cloud of lower chlorine concentration covered an expanding area along the ground. The wind was unusually calm and slowly pushed this cloud in the uphill direction, toward the command post and toward more populated areas to the east and west. Indications are that the gas cloud simply expanded under the cool and dry atmospheric conditions and moved slowly over the general area, while the amount of vaporized chlorine in the atmosphere continued to increase.

Weather Conditions:

Temperature	65 degrees Fahrenheit
Relative Humidity	15 percent
Wind	0-5 mph, variable direction

As time progressed it became evident that the chlorine cloud was continuing to grow. Subsequent investigation showed that the corrosive properties of the liquid had damaged the valve on the discharge side of the transfer pump, allowing a continuous supply of chlorine from the contaminated storage tank to reach the leak. The leak also increased in flow rate, as the corrosive properties of the liquid ate a growing hole in the steel elbow. These factors created a pool of liquid, which allowed increasing quantities of chlorine to vaporize into the atmosphere. Before the leak was stopped an estimated 70 tons of liquefied chlorine gas escaped, making this one of the largest recorded leaks of this type in the United States.

EVACUATION

At approximately 0330 hours conditions began to deteriorate rapidly. The command post and staging area were suddenly enveloped by the gas cloud and had to be evacuated. Although the command post was uphill from the plant, unusual wind conditions allowed the gas cloud to move along the ground and envelope the area without warning. The command post was relocated first to a convenience market parking lot, which also became untenable, and then to a race track parking lot, several miles from the source of the leak. Reports of strong odors in the occupied residential areas and downtown portions of Henderson caused the incident commander to begin evacuations of residents in affected areas.

At approximately 0345 hours a state of emergency was declared by the County Manager, and the Clark County Emergency Operations Center was activated. Preparations were made to provide temporary shelter for large numbers of evacuees, using designated schools and the many hotels and motels in the Las Vegas area. The Red Cross established temporary shelters in predesignated schools, while the Convention and Visitors Bureau began to notify its members of the situation.

The large parking lot at the race track provided an area for assembling resources, in case larger scale evacuations were necessary. Fire, medical, police, and other resources from several jurisdictions were staged and organized to respond to changing conditions. The number of agencies and jurisdictions involved utilized a variety of radio systems and frequencies. This problem was somewhat resolved by the fact that most of the fire service command officers and some of the police officials had cellular phones in their vehicles and these were used to supplement radio communications. No specific serious problem was caused by these communications difficulties though the overall operation was not as smooth as it could have been if integrated radio communications systems had been available.

The Clark County School District made school busses available to the fire department and 50 off-duty firefighters were called in to drive them. Each bus driver was provided with an SCBA, in case a contaminated area was encountered, and two teams were made up with full crews of SCBA-equipped firefighters standing by to take busses into contaminated areas to rescue residents in immediate danger. Police officers were assigned to notify residents in the predicted path of the cloud, while firefighters were assigned to areas where the presence of chlorine could be detected. An estimated 2,400 residents were evacuated from their homes and from local businesses.

Monitoring of the gas cloud was continued by fire department Hazmat teams from Clark County, Las Vegas City, and Nellis Air Force Base, in addition to county and State environmental health personnel. Sampling tubes were used to measure the chlorine concentration at ground level, and the readings were plotted on a map to track and predict movement of the cloud. A police helicopter was assigned to visually track the movement of the cloud.

Most of the readings indicated concentrations that produced a strong odor and respiratory distress but were below levels immediately dangerous to the life and health of most persons. All hospitals in the metropolitan area were alerted to stand by for large numbers of patients, and approximately 250 were examined in emergency rooms. Several police officers, who were exposed while evacuating residents, directing traffic, and performing other functions in the area, were among those patients. The majority of the patients recovered quickly when removed from the contaminated atmosphere. Individuals with asthma and other breathing problems were severely affected by even minor exposure to the gas and accounted for most of the hospital admissions.

St. Rose Dominican Hospital, in downtown Henderson, was located within the affected area, and a decision was made to leave the patients in the building with the air handling system set to recirculate the interior air. Air samples indicated that the ventilation system could maintain a safe interior atmosphere, in spite of the outside conditions. This was felt to be preferable to risking moving the patients outside into the contaminated atmosphere. A retirement home in the direct path of the gas cloud was evacuated.

LEAK CONTROL

As the sun began to rise, the cloud became more visible and its extent could be estimated. Several square miles were blanketed with a greenish haze, sitting just above ground level. It became obvious that the leak was not controlled and that direct action would be needed to stop the flow. It was not known how much liquid had leaked out, but the amount that could potentially escape was now estimated at 100 tons or more. Consultation with plant personnel led to the conclusion that the leak could only be stopped by placing a blank flange in the line between the storage tank and the leaking elbow. An entry team was assembled, including both fire department and plant emergency personnel, to go in and insert the blank flange.

> Note: A blank flange is a solid plate, cut to the appropriate size to fit between the two pieces of a flange pipe connection. To insert the blank flange, the bolts must be removed from the existing flanged connection and the plate inserted between the two halves. The bolts are then reinstalled and tightened.

The entry team dressed in Level B protective clothing and SCBA approached the area of the leak. They found that the hole in the pipe had expanded to approximately one-inch diameter and was producing a steady stream of liquid chlorine. The liquid had formed a pool from which the chlorine was vaporizing rapidly.

To install the blank flange it was necessary to work within two feet of the actual leak, under the spray of the leaking liquid. Plant employees performed the actual installation of the blank flange, backed-up by fire department members. The first attempt, at approximately 0630 hours, proved unsuccessful when the highly corrosive liquid damaged the steel plate before the bolts could be tightened. A second attempt was successful, at another flange location, using a Teflon-coated steel plate. The leak was stopped by approximately 0730 hours.

With the leak stopped, the problem began to diminish. The remaining cloud on the ground vaporized quickly and the vapor cloud began to dissipate under improving weather conditions. Over the next three hours the heat of the sun warmed the ground and provided buoyancy for the cloud, while a more normal wind condition helped to move it away from the populated area. By 1000 hours the cloud had dispersed completely and evacuees were allowed to return to their homes and businesses.

For the next 24 hours there was a continuing concern that more of the stored chlorine might have been contaminated with water and the corrosive mixture could be eating its way out of additional tanks and piping. The Clark County Fire Department Hazmat Team continued to stand by until sampling of the remaining storage concluded that the contaminated chlorine was contained in one storage tank and most of its contents had leaked out. Preparations were made to disassemble the heat exchanger where the problem was believed to have originated.

LESSONS LEARNED

1. **The problems associated with a high-risk occupancy in one jurisdiction, creating a problem in a different jurisdiction, present obvious challenges for emergency planning response agencies.**

   In this case the responding agencies worked well together, but the deficiencies of the regulatory and planning processes were a major focus of attention after the incident.

2. **The Incident Command System (ICS) proved to be extremely effective in this incident, particularly in coordinating the efforts of several different agencies at the scene.**

   The ability to assign major responsibilities to command officers from different fire departments, without any problems, is evidence that the personnel are trained and prepared to operate effectively.

3. **The lack of effective radio communications between agencies was a problem at this incident.**

   Cellular telephones were used very effectively to supplement public safety radio capabilities and proved to be reliable under these circumstances. In other situations cellular telephone service has been compromised by the number of persons trying to use the systems under emergency conditions and particularly the heavy use of the systems by news media personnel. The location and time of day may have been key factors in making the cellular network responsive in this incident. Note: Centel Cellular will block off communications allowing only emergency personnel phones to work if the system starts to overload.

4. **The decision between evacuating residents and warning them to remain indoors, with windows and outside air inlets closed, is often critical.**

   In this case, it was considered more practical to keep patients inside the hospital than to expose them to the outside atmosphere. This took into consideration the susceptibility of the patients to chlorine exposure, the ability to exclude outside air from the ventilation system and the availability of medical personnel and equipment inside the hospital. Some area residents reported that they were notified to evacuate and were exposed to the chlorine cloud while waiting for busses to pick them up. The risk of exposure during evacuation may have been greater than the risk if they had remained indoors.

5. **The use of busses operated by fire department personnel is a practical means to evacuate residents.**

   It is more feasible to have firefighters drive busses than to train bus drivers to use SCBA. It is equally difficult for firefighters using SCBA to convince residents to expose themselves to the outside atmosphere in order to evacuate, unless the residents are already in distress. The contingency plan, sending a crew of SCBA-equipped firefighters on a bus to enter and evacuate an area in immediate danger, is a practical innovation.

6. **Police officers, who are not provided with or trained to use SCBA, were effective in evacuating areas ahead of the contamination, but could not function in the contaminated areas.**

   Several police officers who were assigned to traffic control or to assist with evacuation were exposed to the chlorine cloud and transported themselves to medical facilities for evaluation in the later stages of the incident.

7. **Due to the relatively low concentration of chlorine in the gas cloud, in this case, the predominant medical condition was limited to short duration respiratory irritation.**

Chlorine is detectable by odor at very low concentrations and is a respiratory irritant between 3 and 30 ppm. Individuals with chronic respiratory problems, such as asthma, were quickly affected and accounted for most of the hospital admissions.

8. **It proved to be extremely difficult to determine the size, shape, and movement of the chlorine cloud.**

Helicopter observation was a valuable asset, particularly with increasing daylight. Ground sampling over large areas is difficult to coordinate and requires careful mapping to be effective. An attempt was made to predict dispersion of the chlorine using CAMEO (a computer model program), but complicated factors of terrain, slope, temperature, wind velocity, relative humidity, and an unknown rate of release made predictions extremely difficult.

9. **The application of water to the vapor cloud was considered in this situation to accelerate the evaporation of the pooled liquid.**

Conventional wisdom suggests that massive applications of water spray could absorb chlorine from the air, resulting in a dilute liquid solution. Chlorine has a low rate of solubility in water, and, with a large leak, there is a concern that applying less-than-sufficient volumes of water would create a corrosive fog. Applying water to a container of liquefied chlorine could heat the contents to their boiling temperature and cause the container to rupture. It is often difficult or impossible to estimate the flow and rate of vaporization from a leak to make such determinations.

10. **The delay in notification of the fire department and other agencies indicates a problem with plant personnel and the established standard operating procedures at the facility.**

A review of communications tapes reveals that no call was received by the police or fire departments for this incident from the facility. A private-sector ambulance provider had been requested to respond to transport plant employees who had been exposed to the chlorine gas.

11. **While emergency procedures had been planned for the chlorine facility itself, there was no specific plan for notification or evacuation in the event of a chlorine leak or other emergency extending beyond the property line.**

Due to the risk created by the chemical industries in the area, the need for emergency warning systems should be evaluated. This level of planning requires both private- and public-sector participation.

12. **The fact that the incident occurred at the facility where the CHLOREP Team equipment was stored caused unusual problems.**

Most of the equipment that could have most valuable in trying to secure the leak could not be reached because it was in the Hazmat area. This included the SCBA units normally used by the team members, which are of a different type from those used by area fire departments. Another chemical plant in the area was able to provide the needed SCBAs for the plant members on the entry team.

APPENDIX A

Chlorine Leak Incident Area Map, Pioneer Chlor Alkali Facility, and Diagram of Leak Location

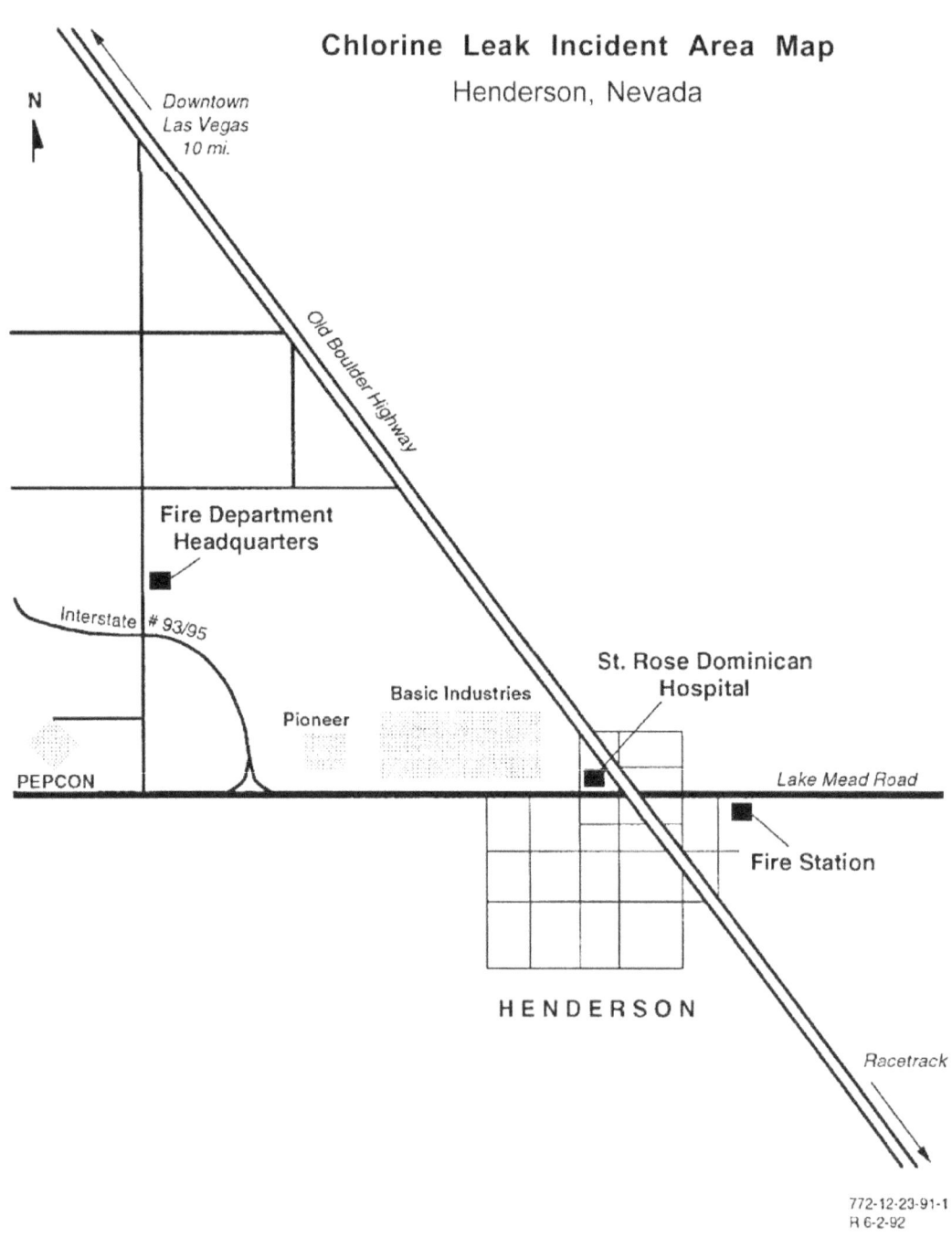

Chlorine Leak Incident Area Map
Henderson, Nevada

772-12-23-91-1
R 6-2-92

10

Appendix A (continued)

Pioneer Chlor Alkali Facility

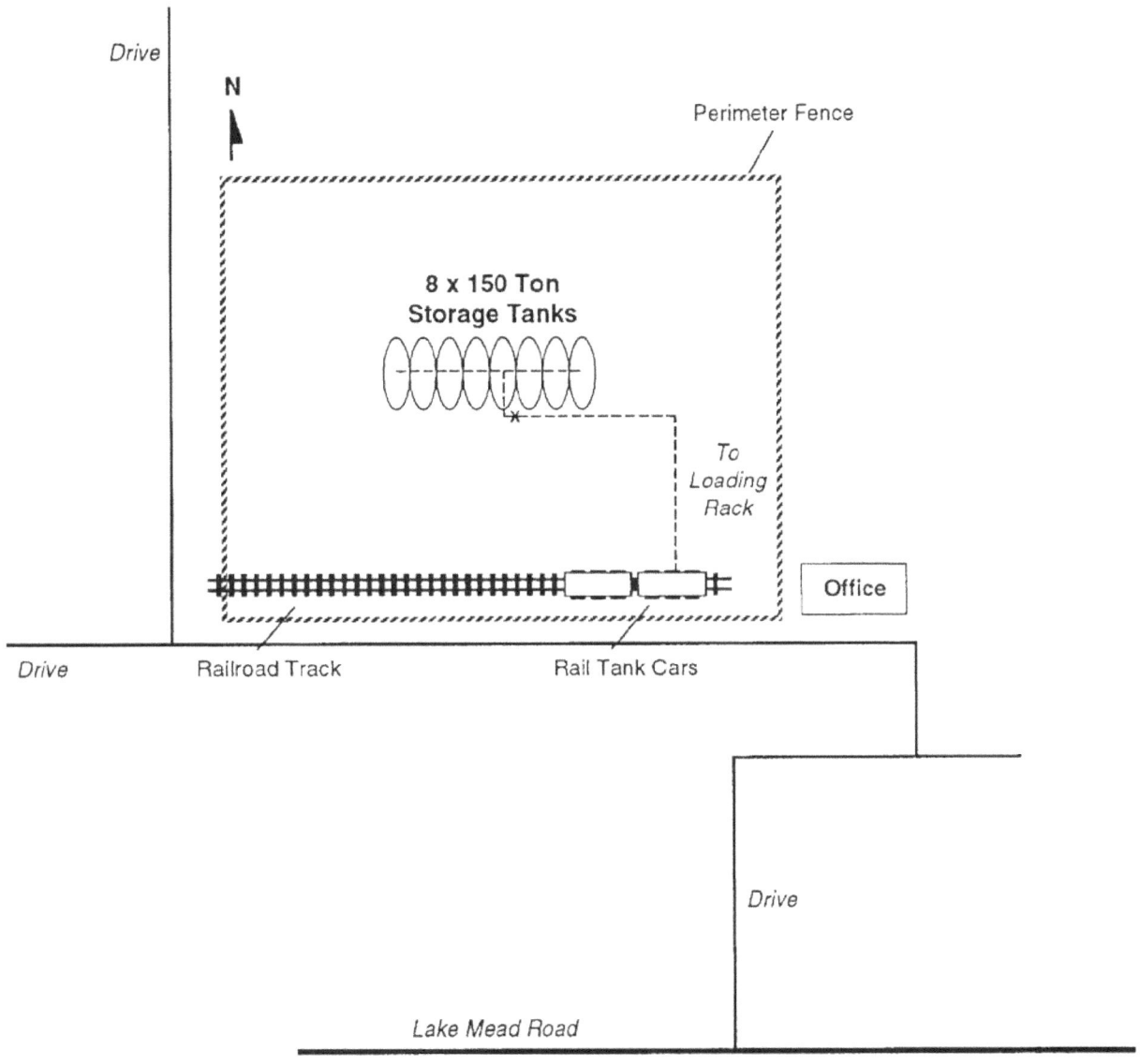

772-12-23-91-2
R6-2-92

Appendix A (continued)

Diagram of Leak Location

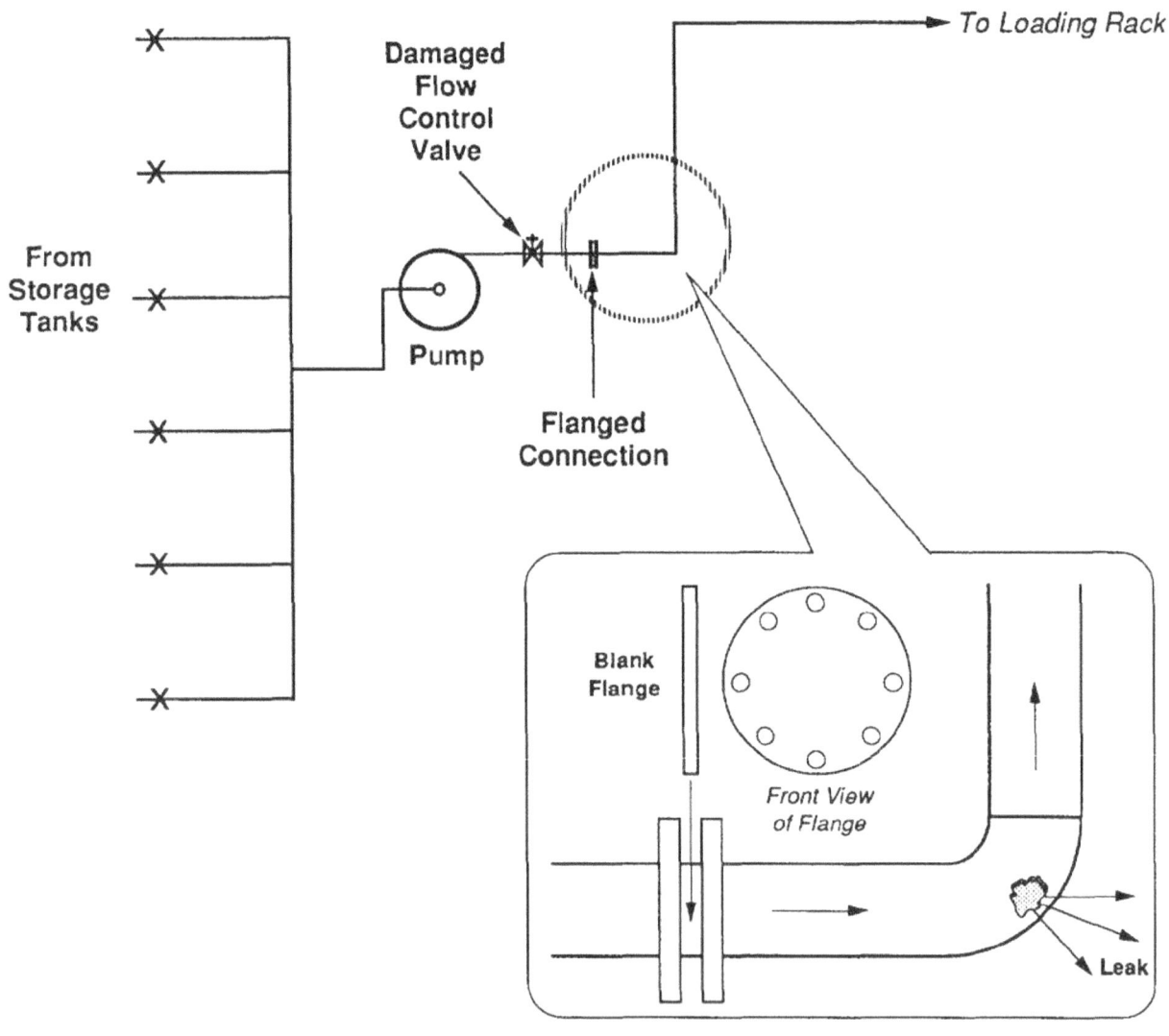

772-12-23-91-3

APPENDIX B

Technical Articles on Chlorine and Dealing with Chlorine Emergencies

Appendix B (continued)

Chemical Data Notebook Series:

Chlorine

BY FRANK L. FIRE

According to an unreleased report by the U.S. Environmental Protection Agency (EPA) concerning hazardous material incidents from 1980 through 1985, chlorine caused more deaths and injuries than any other toxic chemical involved in these accidents. Out of some 6,928 incidents reviewed by the EPA, chlorine was involved in approximately 665 accidental releases and caused 135 deaths and nearly 1,500 injuries.

PROPERTIES AND APPEARANCE

Chlorine is a member of the family of elements known as halogens (other members are fluorine, bromine, iodine, and astatine, which is radioactive and very rare). A yellowish-greenish gas that is 2.45 times heavier than air and very slightly soluble in water, chlorine does not exist in the elemental form, but in molecular form. The chemical symbol for chlorine is Cl, but when it is generated, it exists as the diatomic molecule, Cl_2.

Like many other gases, chlorine is usually liquified for transportation and storage. This is done for economic reasons, since you can get so much more material into a given volume as a liquid rather than as a gas or vapor. For instance, chlorine produces 457 cubic feet of gas for every cubic foot of liquid. This means, of course, that you can store or transport 457 times more chlorine as a liquid than as a gas.

Liquid chlorine (like the gas) has a greenish-yellow tint, and a specific gravity of 1.56 (as opposed to its vapor density of 2.45). Of course, whenever liquid chlorine is exposed to air, it begins to boil away since its boiling 'point is about -30°F.

The chemical properties of chlorine make it an oxidizer, a corro-

sive, and an irritant (as DOT might classify it) or toxic (as IMO classifies it with a 2.3 designation at the bottom of a DOT placard). All these features make chlorine a highly reactive material, which is the reason that chlorine is so important to industry, being used in the manufacture of a great variety of chemicals and materials, among which are bleaches, plastics, rubber, dyes and pigments, pulp, paper, refrigerant gases, fire extinguishing agents, disinfectants, and in the production and/or processing of innumerable specific organic and inorganic chemicals.

Chlorine may also be found in large quantities in water treatment and sewage treatment plants. Chlorinated hydrocarbons (which means that chlorine has been chemically attached to a hydrocarbon by substituting it for a hydrogen atom) are very valuable to industry as solvents, de-greasers, and other important commercial uses, so chlorine may be found in any industrial plant that manufactures these materials.

Chlorine has one synonym, bertholite, but may sometimes be referred to as molecular chlorine or liquid chlorine.

IDENIIFICATION

Chlorine containers can be identified by the use of the Department of Transportation (DOT) nonflammable gas placard with the UN designation 1017 in the center of the placard or next to the placard. Its Standard Transportation

FRANK L. FIRE is the director of marketing for Americhem Inc. in Cuyahoga Falls OH. A chemistry of hazardous materials instructor at the University of Akron, he is also an adjunct instructor of hazardous materials at the National Fire Academy. Mr. Fire is the author of A COMMON SENSE APPROACH TO HAZARDOUS MATERIALS a new book published by Fire Engineering Books

Commodity Code (STCC) number is 4904120, and its National Fire Protection Association (NFPA) rating (704) is 3-0-0-OXY. DOT required labels are non-flammable gas and poison.

HAZARDS

Chemical actions and reactions

Although chlorine does not burn, which is the reason for the DOT designation as a non-flammable gas, like oxygen, chlorine does support combustion. As a matter of fact, chlorine is almost as efficient an oxidizer as oxygen. This means that any ordinary combustible or very flammable material may become explosive when mixed with chlorine. Therefore, all combustible materials, particularly organic substances, and all powdered metals and many metal compounds must be kept separated from chlorine.

If chlorine is released anywhere near a fire incident, efforts must be made to keep the chlorine gas from reaching the fire. Since chlorine is a very powerful oxidizing agent, it will intensify the fire to the point where no ordinary firefighting effort will be able to control it-similar to the addition of pure oxygen to the fire.

The fact that gaseous chlorine is so heavy (2.45 times as heavy as air) means that the gas "hangs together" and flows along the ground, seeking low spots in the terrain. Obviously, these areas are extremely dangerous because of the concentration of chlorine. This hazard can exist quite far from the initial incident, and, depending on the size of the release, evacuation distances downwind may have to be extended one to two miles or more.

An interesting method of chlo-

Reprinted with permission from FIRE ENGINEERING

Appendix B (continued)

rine detection in low-lying areas, and in basements or other confined areas into which chlorine might have flowed and become trapped because of its high vapor density, is the use of a rag soaked with ammonium hydroxide, tied to the end of a long pole. If this rag is waved around in an atmosphere containing chlorine, a white cloud will form wherever the rag contacts chlorine. However, although this detection method will warn of the presence of chlorine, it will not indicate the *level* of concentration.

If you suspect that chlorine is present, keep your mask on. The use of gas sampling and detection devices will accurately measure the presence of chlorine at levels that are very low, but still harmful.

Driving into a cloud of chlorine is dangerous to the vehicle as well as to the occupants. Being a strong oxidizer, the chlorine will support the combustion of gasoline or diesel fuel, and there can be severe damage (corrosion) to the engine if chlorine is pulled through with air, making the engine race as if at full throttle.

In addition to its oxidizing power, chlorine is a very strong corrosive, especially where large quantities come in contact with water. Chlorine will react with almost all metals at elevated temperatures, and some metals, like copper, may spontaneously ignite in the presence of chlorine.

Personal

Needless to say, chlorine is very corrosive to skin and eyes, and contact with the liquid or gas must be avoided at all costs.

If being an oxidizer and corrosive is not enough to make chlorine highly hazardous (and it is), it is also toxic when present in sufficient quantities. The TLV (threshold limit value) for chlorine is 1 ppm and the short-term exposure limit (STEL) is 3 ppm for 15 minutes. You should be able to detect the presence of chlorine around the 3 ppm level (perhaps even as low as 0.02 ppm) as a pungent, choking, irritating odor, described as acrid. It may also resemble the odor of household bleach (which contains chlorine).

Irritation of the eyes, mucous membranes, and respiratory tract

may occur at concentrations between 3 ppm and 15 ppm. Exposures at 15 ppm will cause immediate irritation of the throat, while levels of 50 ppm are dangerous, of-

---

Glossary

Atom - The smallest part of an element that can still be identified as the element.

Boiling point - The temperature at which the vapor pressure of a liquid just equals atmospheric pressure.

Diatomic - Made up of two atoms, as in a diatomic molecule.

Element - A pure substance that cannot be broken down into simpler substances by chemical means.

Gas - A state of matter defined as a fluid with a vapor pressure of 40 psia at $100°F$.

Molecule - A chemical combination of two or more atoms, either of the same or different elements. The smallest particle of a compound that can still be identified as the compound.

Specific gravity - The weight of a solid or liquid as compared to the weight of an equal volume of water.

STEL - Short-term exposure limit. The maximum amount of material to which a person may be exposed over a period of time without harmful effects.

TLV - Threshold limit value. The amount of a substance to which an average person in average health may be exposed in a 40-hour work week without harm. The values may be averaged over time, and the TLV may also be referred to as TWA or time weighted average.

Vapor density - The relative density of a vapor or gas as compared to dry air.

Vapor pressure - The pressure exerted by vapor on the sides of a container at equilibrium. Equilibrium is reached when the vapor pressure of the vapor in the container has stabilized.

---

ten resulting in severe breathing difficulties, and exposure to chlorine concentrations of 1,000 ppm for even a very brief period may be fatal.

Since chlorine is so irritating, exposure to very high concentrations is rare, unless the exposed person is unable to leave the area. Initial irritation of eyes and the mucous membrane of the nose and throat is followed by coughing and a constriction of the chest, accompanied by a feeling of suffocation and pain. Pulmonary edema follows in severe exposures.

Victims of severe inhalation problems must get fresh air immediately, artificial respiration if breathing has stopped, and medical attention as soon as possible. (For those applying artificial respiration, beware of the chlorine in the victim's airways.)

For those victims who have contacted the liquid, all contaminated clothing must be removed, and all affected body parts must be washed with large amounts of water. Medical attention must be given immediately.

PROTECTIVE CLOTHING

Protective clothing is required if there is any possibility of contact with chlorine. This means wearing positive pressure self-contained breathing apparatus (SCBA), full faceshields, rubber boots, as well as gloves and clothing that is impervious to chlorine. Impervious materials include polyvinyl chloride, chlorinated polyethylene, Viton, and neoprene. Fully-encapsulating suits may be required in some instances.

Gas masks with chlorine cartridges or chlorine cartridge respirators with full facepieces are satisfactory for low concentrations (25 ppm or less), but should not be used if there is the slightest chance that the concentration may be higher.

HANDLING

Spills

Remember that spilled/released liquid chlorine will be in an environment that is somewhat warmer than the chlorine itself, and this will promote boiling and rapid generation of chlorine gas. Be

Reprinted with permission from FIRE ENGINEERING

Appendix B (continued)

aware of the generation of this gaseous chlorine and the direction it will travel.

Water fog may be used to direct the movement of gases and will even absorb some of them. In all cases, however, be careful that water is not added to the spill, since this will increase the generation of gases. Also be careful to account for runoff, as this will contaminate water supplies.

NOTE: Although chlorine is used to purify water, the amount that ends up in the drinking water supply is carefully controlled, and excessive amounts can be very harmful.

In the case of a spill on land, containment procedures used for other liquids may be utilized, even though chlorine will be rapidly evaporating since the ambient temperature will probably be above chlorine's boiling point. These techniques include digging a pit, building dikes, or digging trenches. Sand or soil may be used as diking material. In each case, however, liquid chlorine may seep into the soil, spreading contamination. Usually after an incident has ended, several inches (or feet) of contaminated soil may have to be removed.

If it is deemed necessary to retard the release of chlorine gas from the exposed liquid, a fluoro-protein foam or a special chlorine foam may be applied to the surface of the spill. This procedure should be considered only as a short-term solution, however, since the foam will break down and the generation of gases will resume.

If the release is the result of a ruptured container and no pit has been dug or no dike has been constructed, a corrosion-resistant pump may be used to pump the liquid back into the leaking container or into chlorine-resistant containers (made of plastic, glass, or the same metal as the original container). These are short-term solutions, producing semi-closed systems to hold the chlorine until an adequate/proper container can be secured.

Again, any exposed chlorine liquid will be boiling and generating chlorine gas, so anyone working on the containment procedure must be properly protected and

Reprinted with permission from FIRE ENGINEERING

Appendix B (continued)

evacuation downwind must be a prime consideration.

The use of chlorine kits is recommended by shippers to handle small leaks from various size chlorine containers. The kits, labeled A, B, and C, are for containers ranging from one ton to bulk storage railcars.

All efforts must be made to keep the liquid from entering a sewer or waterway, as this will cause the intimate mixing of organic material (fuel) with the liquid chlorine (oxidizer). All that is needed to complete the fire triangle is an energy/ignition source, which, if found, can cause a spectacular underground explosion that could affect an entire city.

Entrance of liquid chlorine into a waterway will also cause serious problems to the water downstream. Care must be made to warn *all* users of the water, industrial as well as municipal. In all cases, the operators of the sewage treatment plant through which the contaminated water may pass must be notified.

In some cases, mixing activated charcoal into the contaminated water will cause adsorption of the chlorine onto the charcoal, which could then be removed from the water by screening it out. In every instance, however, notify water users downstream.

Neutralization

Some references call for neutralizing chlorine spills, but do not specify the proper neutralization agent. One possibility is absorption with fly ash or cement powder and the addition of caustic soda. As in any situation where chemicals will be added to accidentally released substances to neutralize them, care must be taken not to cause the situation to worsen.

When preplanning for the possibility of a chemical release, qualified experts should be consulted to determine safe neutralization techniques, and, if possible, retained on call in the event of an emergency.

Consideration should be given to trying to divert contaminated water until it can be properly heated to eliminate the chlorine. This may require damming and / or diking the waterway, and diverting the water through some land-based decontamination station. It may even be possible to divert the contaminated water through a sewage treatment plant, but permission must be obtained from the sewage plant operators before this tactic can be used.

SUMMARY

In any release, all emergency personnel must be aware of *all* the hazards presented by chlorine. Since chlorine boils at -30°F, that must be the maximum temperature of the liquid, and any water applied to it will be at least 63° warmer, causing an increase in the generation of gaseous chlorine. This will increase the intensity of the fire *and* the amount of toxic gases in the air.

Also be aware of the possibility of uninvolved chlorine containers exploding whenever they are subjected to extreme heat. ■

Reprinted with permission from FIRE ENGINEERING

Appendix B (continued)

Fire Science

DEALING WITH CHLORINE EMERGENCIES

CHRISTOPHER KELLY

Los Angeles firefighters battling a greater-alarm fire in a lithography plant were unaware that 15 cylinders of chlorine were stored illegally in the building. When heat melted the safety plugs on the 150-pound cylinders, poisonous chlorine gas affected 44 firefighters, some severely.

Five months later, Los Angeles firefighters were fighting another greater-alarm in Leslie's Pool Mart, when water mixing with chlorinated powder in burning plastic and cardboard containers resulted in a gigantic cloud of chlorine gas. Some 5,000 people were evacuated, and more than two dozen firefighters suffered from chlorine inhalation.

Kits and gear recommended by the CI.

These fires illustrate the problems with chlorine as a liquefied gas under pressure, and as a calcium or sodium hypochlorite and other chlorine-containing compounds. Except under unusual circumstances, chlorine is neither explosive nor flammable, although it will support combustion. Chlorine has traditionally been a scare word in the fire service. It need not be. With knowledge and training, chlorine emergencies can be handled with relative safety.

Because chlorine is one of the most widely-used chemicals, firefighters can encounter it in both industrial districts and residential neighborhoods. While chlorine is popularly associated with water purification and laundry bleaches, its major users are producers of plastics, synthetic fibers, solvents, and paper. With backyard swimming-pools becoming commonplace throughout the United States, quantities of water-purifying hypochlorite in liquid, granular, or tablet form are stored under various trade names at suburban hardware and pool-supply stores.

Cylinders of chlorine are loaded onto truck.

Chlorine's pungent, abrasive stink is unforgettable to firefighters exposed to it. This odor is even detectable at low levels of concentration when chlorine is colorless. Higher concentrations of gas, which can be fatal, produce a greenish-yellow vapor cloud. The chlor-alkali industry, however, has an excellent safety record, and its Chlorine Institute (CI) in New York City provides guidelines for firefighters.

Chlorine emergencies often involve pressurized cylinders containing the liquefied gas. There are three sizes: 100 and 150 pound, and ton.

Chlorine is also transported in railroad tank-cars carrying up to 90 tons,

trucks, which can carry 15 to 20 tons, and barges with capacities of up to 1100 tons. Regardless of size, cylinder valves and fittings are basically similar through industry-wide standardization.

CI emergency kits containing tools and other devices for capping or closing leaking valves and sealing sidewall leaks are frequently kept by users of chlorine. Chlorine-hauling tank trucks are required to carry them in the cabs. Some fire departments, notably Baltimore and Los Angeles, own their own.

Ton container being loaded onto barge.

Task Force Commander Ralph E. Davis of the Los Angeles City Fire Department recently compiled lists of locations of these kits throughout the city. The CI recommends that these emergency kits (A for cylinders, B for containers, and C for tank trucks and rail cars) be made available in an emergency to firefighters or other qualified personnel. There are now 5800 kits in locations around the country. Pamphlet 35, available from the CI, gives their exact whereabouts.

Tank-truck emergencies are fought with kit C.

The CI recommends that: (1) firefighters approach chlorine emergencies from an upwind direction while wearing full protective clothing, including

Firehouse/December 1977

Reprinted with permission from FIREHOUSE Magzine

Appendix B (continued)

self-contained breathing apparatus, (2) only personnel required to remedy the problem enter the contaminated area, (3) for safety, firefighters should work in pairs, (4) as an added precaution, a lifeline be attached to each firefighter.

Evacuation should be pre-planned. Chlorine is two and a half times heavier than air, and therefore tends to cling to the ground, sometimes settling in low spots such as basements, elevator shafts, and pits. Portable fans or blowers commonly carried on apparatus can be helpful in dissipating the gas.

The ability of one cubic-inch of liquid chlorine to expand into about 460 cubic-inches of gas indicates the potentially gigantic vapor cloud that can be formed. Research has shown that a ruptured 55-ton tank car of chlorine can result in the instantaneous release of about 25 percent of the liquid, or about 25,000 pounds of chlorine. Thereafter, 10 pounds of chlorine per second will be continuously emitted as the liquid evaporates due to heat picked up from the surrounding environment.

Chlorine emergencies are generally of two types:

(1) *Leaking Containers.* Immediate action must be taken because chlorine leaks become progressively worse. The CI emphasizes that water must *never* be used, because the corrosive combination of water and chlorine increases leaking. The best method for locating leaks is with a swab dipped in commercial aqua ammonia, often kept at large chlorine installations. The mix of ammonia and chlorine will result in puffs of white vapor, itself highly toxic, from the leak. Equipment and piping leaks can best be stopped by shutting off the valve with a crescent wrench. Puncture leaks can be reduced by turning the cylinder so that gas escapes instead of liquid, since the weight of the gas that would escape would be one-fifteenth that of the liquid chlorine that would escape from a comparable puncture.

Leaking tank cars or trucks create other problems. "It is generally advisable to keep the vehicle or tank car moving until open country is reached in order to disperse the gas and minimize the hazards of its escape," says the CI.

(2) *Overheating Under Fire Conditions.* Chlorine problems can escalate rapidly because the fusible safety plugs will melt at 158 to 165 degrees Fahrenheit. The CI recommends that cylinders, ton containers, tank cars, and trucks be removed immediately from the danger area. If the cylinders cannot be moved and are not leaking, cooling fog streams should be applied continuously until the fire is out.

Fire poses another threat at truck crashes or tank-car derailments. If the truck is carrying cylinders, those intact should be removed. When a truck or tank car is leaking and cannot be moved, the area must be evacuated.

Hypochlorite and other chlorine-containing products present a serious challenge to firefighters. There is disagreement over the best way to handle fires involving hypochlorite. "It's a damned if you do, damned if you don't situation," says LAFD Battalion Chief Leo K. Najarian, a hazardous-chemicals expert. "Most chemical compounds associated with servicing swimming pools are formulated to produce chlorine on contact with water," says Najarian. "In addition to liberating chlorine, many of them also release oxygen and are therefore classified chemically as oxidizers or supporters of combustion. "Companion chemicals relating to pool-service are mineral-acid compounds such as muriatic (hydrochloric) acid which are used to maintain the proper pH balance. When these pool chemicals are exposed to heat, water, and contamination with other chemicals, and, come in contact with combustibles, the result can be fire, explosion, and release of extremely toxic gases such as chlorine and products of combustion."

Najarian says that fire hazard is acute in establishments supplying these chemicals, which are often packaged in combustible plastic and cardboard containers. "Firefighting activities in this type of occupancy are extremely critical. The application of water for extinguishment purposes can create greater problems than those posed by the fire, due to the resulting production of toxic gases. On the other hand, if water is not applied, the gases will still exist, but not in the amounts that the water would cause. Without water, the possibility of an explosion occurring from an uncontrolled fire involving large amounts of these oxidizers becomes a definite possibility.

There is no ideal solution. One approach is a fast, hard-hitting attack with master streams boring into the fire, while firefighters try to remove chemicals that are not involved, or to avoid wetting them. "This is easier said than done," admits Najarian. "It would appear that exposures of the public to this hazard could best be avoided through stricter zoning laws that would require more isolation of the chemicals themselves and at greater distances from inhabited areas."

Chlorine-exposure victims are most likely to suffer eye irritation, coughing, and labored breathing. In extreme cases, death by suffocation results. There is no known antidote for inhalation of chlorine gas. The best treatment is 100-percent oxygen at atmospheric pressure for no longer than one hour at a time. Full recovery is certain in virtually all cases. Skin or eye contact with chlorine will result in burns. Recommended treatment is thorough flusing of the affected area with water.

The CI offers free round-the-clock emergency assistance for potential or actual emergencies through CHLOREP (*Chlor*ine *E*mergency *P*lan) consisting of 60 teams at plant locations in the United States and Canada. CHLOREP experts in the 48 contiguous states can be reached by calling CHEMTREC (*Chem*ical *T*ransportation *E*mergency *C*enter) in Washington DC, at this toll-free number, 800-424-9300. Canadian help will be provided through TEAP (Transportation Emergency Assistance Plan). The Canadian- phone number varies by region.

Planning is the key to minimizing chlorine problems. (1) Know where chlorine is being used and stored in your first-alarm district. (2) See that all chlorine transportation and storage facilities are properly labelled. (3) Make certain both you and the users have an emergency plan.

With planning and education, chlorine emergencies should be no cause for panic.

Reprinted with permission from FIREHOUSE magazine

Appendix B (continued)

CHLORINE

1017 / **2**   **3** | **0** | **0** Oxy

Poison gas, non-flammable, **oxidizer**   UN: 2.3   STCC: 49-041-20

FORMULA: CL$_2$

OTHER NAMES: Bertholite (Former military poison gas)

PHYSICAL PROPERTIES: Vapor Density: 2.45 (much heavier than air) Specific Gravity: 1.56 (heavier than water) Boiling Point: −30 degrees Fahrenheit (Water will cause boiling) Ignition Temperature: Non-flammable Water Solubility: Slight (0.65g/100g) Color: Greenish-yellow (small amount of gas may be invisible) Odor: Pungent like bleach, choking, irritating Expansion Factor: 457: 1

NOTES:

POISON, OXIDIZER, CORROSIVE, IRRITANT: Can make metals burn and organic materials, including rubber and plastic, explode. Chlorine has caused more deaths and injuries than any other hazardous material. Chlorine is shipped as a liquefied compressed gas. (Treat as a cryogenic.) Most containers are fused at 160 degrees Fahrenheit, but even they can explode in heat.

— — — — — — — — — FOLD — — — — — — — — —

HEALTH: Poison, irritant. (Must wear positive pressure SCBA, protective gear.) One thousand parts per million causes severe breathing difficulty. Fifteen parts per million causes throat irritation. Three ppm irritates eyes and mucous membranes.

FLAMMABILITY: Chlorine is non-flammable, but it is an oxidizer and will intensify fires. It may make flammables explode (even metals like copper and aluminum may burn when heated in the presence of chlorine).

REACTIVITY: Strong oxidizer, irritant (see above). Reacts with water to form dilute, but corrosive hypochlorous acid, which decomposes on standing to yield small amounts of chlorine, oxygen and chloric acid. Can form explosive mixture with ammonia. The corrosive action of chlorine and water will make leaks worse.

DOT RESPONSE GUIDE 20 EXTINGUISHING AGENT: Does not burn

SPILLS: Vapor: Shut off valve or plug. Ventilate if indoors. Liquid: Shut off or plug. Dam, dike or pit. Ventilate if indoors. Use A kit for 100-pound cylinders; B kit for 1-ton cylinders; C kit for tank cars. Do not immerse containers in water.

PROTECTIVE EQUIPMENT: For small gas leaks, positive pressure SCBA, face shields, rubber gloves and boots; seal arm and leg openings with duct tape. (Better to use neoprene, viton, PVC, chlorinated polyethylene suits.)

FIRST AID: (Effects may be delayed.) Decontaminate before placing in the confines of an ambulance. Remove to fresh air, give oxygen, remove contaminated clothing, wash with water, transport.

EVACUATION: Small leak: 140 feet in all directions. Large spill: 0.7 by 1.0 miles. Cylinder in fire: 1500 feet in all directions. Bulk container in fire: 2500 feet in all directions.

"CAUTION: CONDITIONS MAY PRESENT NEW DANGERS. USE ONLY APPROVED LOCAL PROCEDURES."

Firehouse/April 1988
Reprinted with permission from FIREHOUSE Magazine

Appendix B (continued)

HOW TO GET HELP FOR CHLORINE EMERGENCIES

CHLOREP, the **CHLORine Emergency Plan,** has been organized by the Chlorine Institute to advise and assist at any potential or actual emergency involving chlorine gas. It operates on a 24hour. 7-day-a-week basis from 60 plant locations in the United States and Canada

CHLORINE USERS should post suppliers' emergency numbers near point of use, and call them if an emergency occurs. If supplier cannot be reached, call the number below for your location.

EMERGENCY SERVICES handling a chlorine emergency should call the appropriate phone number below.

In the **UNITED STATES,** summon help through CHEM-TREC. the Chemical Transportation Emergency Center at the Manufacturing Chemists Assn. in Washington. D.C.:

48 contiguous states (toll free)   800-424-9300
If "800" number cannot be reached from your phone, call the "202" number instead.

Alaska and Hawaii   202-483-7616
(telephone advice only)
District of Columbia   483-7616

In **CANADA,** summon help through any regional control centre of TEAP. the Transportation Emergency Assistance Plan of the Canadian Chemical Producers Association.

Atlantic Provinces and eastern/central Quebec   819-537-1123 (Shawlnigan, Que.)

Southwestern Quebec   514-373-8330 (Valleyfield, Que.)

Eastern Ontario   613-348-3616 (Maitland, Ont.)

Central Ontario   416-356-8310 (Niagara Falls, Ont.)

Southwestern Ontario   519-339-3711 (Sarnia, Ont.)

Northern/western Ontario   705-682-2881 (Copper Cliff, Ont)

Manitoba, Saskatchewan, and Alberta   403-477-8339 (Edmonton, Alta.)

British Columbia   604-929-3441 (Vancouver, B.C.)

www.ingramcontent.com/pod-product-compliance
Lightning Source LLC
Chambersburg PA
CBHW081419170526
45166CB00010B/3409